D1233146

Animal movement

TONY SEDDON

Facts On File Publications
New York, New York ● Oxford, England

First published in the United States of America by Facts On File, Inc.
460 Park Avenue South, New York, NY 10016.

Library of Congress Catalog Card Number:

88-045747

Designed and produced by BLA Publishing Limited,
East Grinstead, Sussex, England.

A member of the **Ling Kee Group**
LONDON·HONG KONG·TAIPEI·SINGAPORE·NEW YORK

Phototypeset in Britain by BLA Publishing/Composing Operations
Printed and bound in Spain

10 9 8 7 6 5 4 3 2 1

Note to the reader
On page 59 of this book you will find the glossary. This gives brief explanations of
words which may be new to you. Answers to questions are given on page 58.

Contents

Backward and forward

Different animals move in different ways. Some creep, some walk, others paddle through water or fly through the air. At first, these seem like very different kinds of movement. But all moving animals have two things in common. When an animal moves, it pushes backward with its legs, tail, or wings. It is this action of pushing backward that drives the animal forward. Moving animals also keep changing their shape.

▲ Here is a well-known animal famous for being able to run very fast. Can you recognize it? You can see how this animal keeps changing shape as it gallops along. Next time you take a walk on a sunny day, have a good look at your shadow. Does your body also change shape as you walk or run?

 Try taking one step forward with your right foot without pressing against the floor with your left foot.

 The force that moves an animal forward is exactly equal to the force pushing backwards against its surroundings.

▲ An animal on land pushes backwards against the ground and moves forward.

▲ An animal in water pushes backwards against the water and moves forward.

▲ An animal moving in air pushes backwards against the air and moves forward.

"Propellers" are for moving

Animals have devices or "propellers" to make them move. A dog has legs, a fish has fins and a tail, and a honey bee has wings. But sometimes these devices are not so easy to see. For example, what do you think a snail's "propeller" looks like?

▲ A snail's moving parts or "propeller" are underneath on the soft part of its body. Put a snail on a sheet of glass and watch it moving from the other side. Does it follow the two rules of animal movement?

Lots of legs

It is possible to have more than two legs taking turns to swing back and forth. Lots of legs can be arranged in line, one behind the other. Think about a millipede. It may have as many as six hundred legs. We can now see that moving animals work like a machine. A human pushes himself along on two spokes, a dog on four, and an insect on six. A millipede or centipede has a large number of spokes. How many spokes does a spider push itself along with?

Walking wheels

Machines like cars and airplanes move by means of parts that turn round. A wheel turns on its axle. But animals do not have wheels. Instead, they push themselves along by means of rods or levers which move up and down or from side to side. These levers work something like a wheel. A six-spoked wheel like the one in the drawing is like six legs, each ending in a foot. As the wheel goes round, each foot in turn pushes against the ground.

▲ In this wheel only one spoke at a time, with its "foot" attached, meets the ground. After it has pushed backwards, the "foot" moves on. The next "foot" repeats the action. Each "foot" in turn pushes back and so the wheel rolls forwards.

▲ This wheel has only two spokes with "feet" on. Imagine each spoke is hinged at the center of the wheel. After each spoke and its foot has moved backwards, it swings forwards and repeats the action. These are now working like legs.

More about ≫ Walking and running p 34-35 Snails p 25 Millipedes p 25
Spiders p 24-25, 49 Insects p 24, 50-51

Muscle and bone

Most land animals move on legs. Even a snake's ancestors once had legs. However, modern snakes can make do without them. They have found another way to move around. Even so, snakes still have to push against their surroundings but this time with their body. It does not matter how many legs an animal has, they all work in the same way. You use your two legs just like a millipede uses its four hundred or more!

Bones and other hard stuff

Quite a large part of an animal's legs is made of a hard material. In animals with backbones (vertebrates) this substance is bone. In animals such as insects, spiders, and crabs it is called chitin. The hard piece of an animal's legs is part of its skeleton. Vertebrate animals have a skeleton inside the body. In insects and their relatives the skeleton is on the outside. It is an animal's skeleton that forms the levers used for movement.

Joints

A joint is formed where two parts of an animal's skeleton meet and move. In the case of vertebrates the parts are bones. Name two joints in your body.

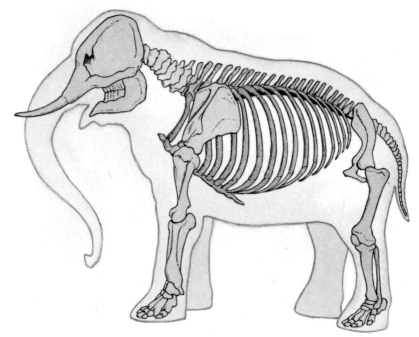

▲ Vertebrate animals have a complicated skeleton made up of over 200 bones. The backbone forms a bridge to help support the animal's weight. Can you see where the leg bones join onto this elephant's skeleton?

▼ A grasshopper's skeleton is found on the outside of its body. The muscles that move the big, back legs of grasshoppers are found inside. You can also see the joints of the leg that allow it to bend.

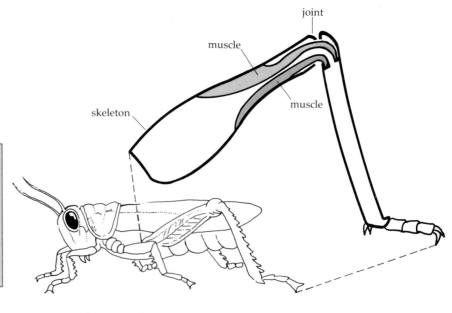

joint

muscle

muscle

skeleton

Muscle power

A car's wheels turn because of the power supplied by the engine. Animals also need power to move their legs, flippers or wings. This power comes from the animal's muscles. The bigger and heavier an animal's body, the bigger and more powerful its muscles need to be.

Pairs of opposites

Muscles that move arms and legs work in pairs. Each one works opposite to the other. A muscle is like a piece of a rubber band. It can be shortened and stretched. When one muscle becomes shorter, its partner becomes longer. This is how muscles move arms and legs, wings, tails and flippers.

▲ These arm wrestlers are showing off their big muscles. The large biceps muscle at the top of their arms has shortened or contracted to bend the arm. What has to happen to straighten the arm again?

Look at the drawing of a tiger. You can see the pair of muscles that work the right back leg. What do you think happens to the leg if:
1 **Muscle A contracts or shortens?**
2 **Muscle B contracts or shortens?**

Muscles and energy

A car cannot move without gasoline in its tank. The gasoline supplies the car with energy and the energy is used to turn the wheels around. Animals also need energy. They get it from their food. If an animal does not get enough food it lacks energy. Without energy its muscles will not work properly and it has difficulty moving around.

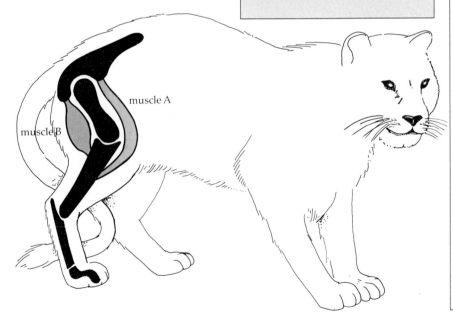

muscle A

muscle B

Limb plans

There are around 50,000 different kinds of vertebrates alive today and they come in all shapes and sizes. The biggest vertebrate, the blue whale, is one billion times heavier than the smallest vertebrate, the dwarf goby fish from the Pacific Ocean.

At first sight, a toad and a lion look very different, and a tortoise does not seem to have much in common with a bird. But in some ways these animals are similar, especially in the underlying structure of their limbs. It is hard to imagine that the leg of a tortoise and the wing of a bird have anything in common. But they do! Both of them are built on the same basic plan.

▼ The vertebrates shown in these pictures move about on structures called limbs. Although these limbs are different in size and shape, their bone structure is very similar. Even a whale's paddle and a penguin's flipper contain the same bones as those in your own arm.

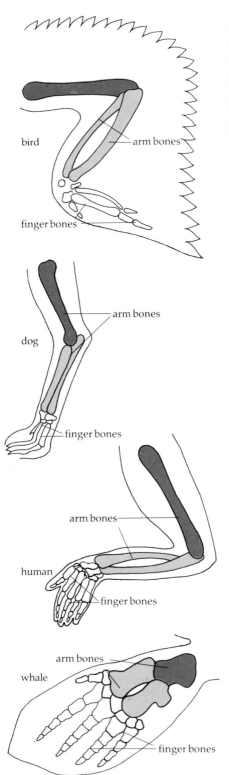

bird
arm bones
finger bones

dog
arm bones
finger bones

human
arm bones
finger bones

whale
arm bones
finger bones

◀ Look at these drawings of some vertebrate limbs. If you think that they are completely different, look again. You can see they have the same plan and the same bones. The main difference is that the individual bones vary in shape and size. This is why each limb looks different on the outside.

▼ This bat has been photographed with the light shining through its wings. It looks like an X-ray picture. You can see all the bones. Can you guess the name of the long thin bones over which the wing membrane is stretched?

A five-fingered limb

The first animals to come onto land from water were fish. This happened millions of years ago. At first the fish dragged themselves along on their stiff front fins. But gradually, over long periods of time, their fins began to change. New bones developed and the fins slowly became legs. All backboned animals have developed from these early land ancestors. This is why the legs of today's vertebrates are so similar. They are built on the same plan. This plan shows one large bone in the upper part of the arm or leg and two more bones in the lower part. The rest of the limb contains ankle or wrist bones and fingers or toes. The usual number of fingers and toes is five. But some animals have fewer than this.

More about ≫ Blue whale p 20-21 Bird wings p 52-53
Bat wings p 54-55

Micro-movement

Even the smallest animals have to move in search of food and safety. But despite their tiny size, many of them follow the same two basic rules of movement. They push and they change shape. However, some microscopic animals are the lazy travelers of the animal world. They are nature's drifters and floaters.

Gone with the wind

Some animals are small enough and sufficiently light in weight to escape from the effect of gravity. They can move in the usual way but sometimes they find other means. They simply float on air. Tiny spiders parachute around the world on threads of silk. Small insects are sometimes carried by the wind high above the ground. These animals are often dropped off in strange places, thousands of miles from where they started. They even land on mountain peaks thousands of feet high. Drifting or floating is a nice, lazy way of traveling.

▲ Young spiders or spiderlings use a method of ballooning to find new homes. Their tiny bodies are so light in weight that gravity has little effect on them. They are almost "weightless," and drift around the world from place to place.

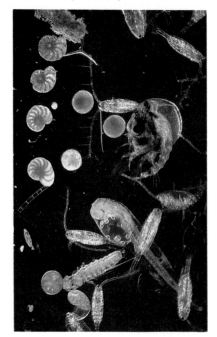

▶ Animal plankton contains tiny animals with strange shapes. The weird projections on their bodies increase the animals' surface area and help them to float more easily.

Drifting along

Many animals that live on the seabed move only occasionally and also very slowly. For example, limpets and barnacles are some of nature's slow pokes! But animals like this are much more active when they are young. But, then, they also look different. Larval forms of animals like shellfish and crabs form the millions of tiny ocean dwellers called plankton. They drift near the surface, often floating for hundreds of miles in the ocean currents. These animals do not make any special effort to move. They are moved by the water. Drifting is one way species spread around the world.

Hairs and whips

Some microscopic animals use very fine hair-like structures to move with. Foot-shaped animals called paramecia are covered in thousands of tiny hairs (cilia) which they use like miniature oars to row themselves through the water. Each cilium sweeps backward like a bendy whip and then forward like a stiff rod. It is similar to the way a tennis player moves his arm when serving. The large number of cilia work in small teams and each is carefully controlled.

Other microscopic aquatic animals have a single whip-like hair or pair of hairs. Each one is called a flagellum. Waves pass down the flagellum to wriggle the animal through the water.

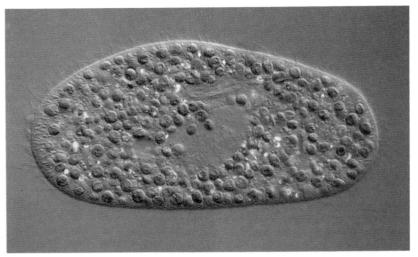

▼ An amoeba moves by squeezing out tiny finger-like parts from the surface of its body. One of these sticks to the muddy bottom. Then the rest of the animal either flows towards or away from it.

▲ A paramecium is a microscopic rower. Although the tiny hairs called cilia seem to move quickly, the animal really moves quite slowly. It probably travels less than 16 feet (five meters) an hour.

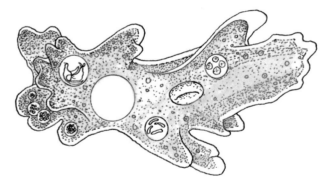

▼ A hydra is a small aquatic animal which somersaults along. You can see it moves "base over tentacles." Some hydra have extra long tentacles which they use to pull themselves up the plant stems.

Liquid engineering

An amoeba is a small, single-celled animal which lives on the bottom of ponds and lakes. It is so small that about twenty individuals would just about cover a pin head. It moves in a very simple way by flowing along like a tiny blob of runny jelly.

More about 〉〉 Spiders p 7, 24-25, 49

Pushing and pulling

Insects and their relatives have a hard outer skeleton to help support them. They also have legs to walk or hop on. But many invertebrate animals are soft-bodied with no hard skeleton and no legs to move about on.

Many worms have their own special way of moving. They depend on a kind of "water skeleton" and two sets of muscles.

The long and the short of it

Although an earthworm looks very different in shape from a cat, it still obeys the same two laws of animal movement. It changes shape and it pushes

▲ Earthworms are tube-shaped animals. They move by stretching and shortening, just like a long piece of rubber or elastic.

against its surroundings. As we shall see, an earthworm also pulls itself along when it moves.

Squeezing and relaxing

An earthworm's body is like a piece of rubber tubing with two sets of muscles. One set of muscles is arranged around the body like a series of tiny rings or hoops. They are called the ring muscles. The other set runs from head to tail like lots of tiny rubber bands. These are called the long muscles. The earthworm also has lots of very small bristles or stiff hairs sticking out underneath its body. The bristles are used to anchor the worm's back and front ends when the different muscles contract.

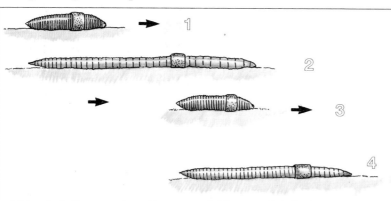

1 The bristles anchor the worm's back end. Its front end is free to move.

2 The ring muscles of the worm contract. The front end is pushed forwards and the worm becomes long and thin.

3 The back bristles unhook and the front bristles anchor the head end. The long muscles contract and pull the back end forward.

4 The bristles anchor the worm's back end again and the process is repeated.

An earthworm moves like a rubber tube which is stretched and then relaxed.

Stuck like a leech

A leech moves in the same way as an earthworm. It has two sets of body muscles which it uses to shorten and lengthen its body. But it does not have bristles. Instead it anchors its body to its surroundings by two suckers, one at each end.

Water works

Starfish move on hundreds of tiny tube feet underneath the body. Water is pumped into tube feet and this makes them swell up like small balloons. Each tube foot ends in a tiny sucker. The suckers help the starfish to get a firm hold on the rocky sea bottom. The starfish can move its tiny swollen legs by means of small muscles attached to them.

1 **The back sucker holds on to the ground but the front sucker is free.**

2 **Now the ring muscles contract or shorten. This squeezes the leech's body and pushes it forward.**

3 **The leech now holds on with the front sucker and lets go with the back one.**

4 **The long muscles now contract and pull the leech's body forward.**

5 **The back sucker holds on again and the front one lets go. The processes in steps 1–4 are now repeated.**

▶ You can see lots of tube feet on the underside of this starfish. It uses them to move along.

Gurgling giants

Giant earthworms over 13 feet (four meters) long are found in South Africa and Australia. Sometimes the worms give themselves away by making gurgling sounds as they burrow under the ground.

More about Insects p 24, 50-51 Insect skeleton p 8
Soft-bodied animals p 24-25 Muscles p 8-9

Moving in water

Water is about 800 times denser than air. Have you ever tried running in the shallow end of a swimming pool? If so, you will have felt the water holding you back. It sticks to the skin and causes friction. Because it drags an animal back as it tries to move through it, this "pulling back" effect is called drag. In order to swim easily through water and to reduce drag, an animal needs the right shape. It needs to be streamlined, like the shapes given below.

▲ A few animals like this nautilus swim by a kind of jet propulsion. They squirt water forward through a tube called a siphon. The strong jet of water squirted out forces the nautilus backward. Dragonfly larvae also jet themselves along in a similiar way. They use this way of moving for high-speed escapes from their predators.

Webbed feet

If you have ever tried to swim with a pair of flippers on your feet, you will know that webbed feet make a great difference to your speed. Many animals have flaps of skin or webbing between their toes. Frogs use their large webbed back feet to push themselves through the water. The duck-billed platypus swims mainly with its large, webbed front feet.

Many water birds have webbed feet. They use them for underwater swimming and for paddling along at the surface of the water.

▼ This turtle uses its large front flippers to pull itself through the water. Penguins have wings that are modified as flippers which they use to "fly" under water at speeds up to 16 miles an hour (25 kilometers an hour). Whales and seals have flippers to help them swim. They also use them for steering, stopping and even back-paddling.

penguin

dolphin

shark

Floating and sinking

Animals are made of denser and heavier stuff than water. Because of this, staying afloat is often a problem. Oil and fat are less dense than water. Animals make use of this fact to help solve their floating problems. The fatty blubber under the skin of seals and whales not only keeps them warm, it also makes their bodies more buoyant. Sharks have oil-filled livers for the same reason.

Many fish have a small balloon called a swim bladder, which works like a tiny buoy. The amount of air in the bladder can be controlled to help make the fish float higher or lower in the water.

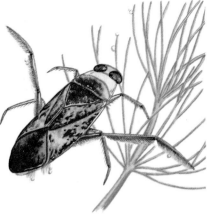

▲ This water boatman has one pair of extra large legs. They work like tiny oars as the insect rows along.

▼ A hippopotamus seems to be able to float and sink as it pleases. It has no difficulty walking on the river bed.

Paddling insects

Many insects that live in water have specially adapted legs to help them swim. Their legs are often covered with rows of tiny hairs. The hairs give the legs a large surface for brushing water aside as they sweep the animal along. The hairs also trap air bubbles, so the animal can take its own air supply with it as it dives.

Rock bottom

Hippopotamuses can actually walk on the river bed like giant underwater tanks, and the nine-banded armadillo does something similar. If a hippo comes to a small stream, it simply carries on walking until it climbs out onto the opposite bank of the stream.

More about >> Swimming p 18-21 Flippers p 20-21 Webbed feet p 40-41
Floating and sinking p 19-20 Insects p 24

Swim like a fish

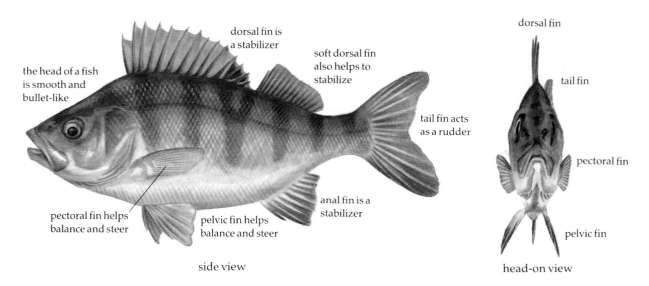

the head of a fish is smooth and bullet-like

dorsal fin is a stabilizer

soft dorsal fin also helps to stabilize

tail fin acts as a rudder

anal fin is a stabilizer

pectoral fin helps balance and steer

pelvic fin helps balance and steer

side view

dorsal fin

tail fin

pectoral fin

pelvic fin

head-on view

We have already seen that a streamlined body shape is important for easy movement through the water. Fish are among the best and fastest swimmers in the animal world. A blue marlin has a perfectly streamlined body and can swim at least ten times faster than an Olympic swimming champion!

▲ This perch has a distinctive streamlined fish shape. The tail fin is used as a propeller and rudder. The other fins help in balancing. The dorsal and ventral fins prevent the body rolling and swinging from side to side. The pectoral and pelvic fins prevent pitching up and down movements. They are also used for steering and stopping.

Powerhouse tail

A fish has a series of muscles on either side of its backbone. These contract in turn and bend the tail from side to side.

Fast-swimming fish are much more muscular than their slower moving relatives. The yellowfin tuna weighs more than 220 lbs (100 kg) and is one of the fastest of all fish. Almost three-quarters of its weight is in its powerhouse back and tail muscles, the main muscles used for swimming.

a fish's tail bends from side-to-side as it swims

Tail bending

When a fish moves its tail from side to side, its body bends due to the pressure of the water. The tail fin pushes backwards and the fish moves forwards. The bigger the tail, the bigger the push and the faster the fish moves.

Wriggle like an eel

Eels seem to wriggle rather than swim. An eel's body is long enough for one part to be moving to the right while another part is moving to the left. It is these waves that push against the water and drive the eel forwards.

Fish out of water

Fish often leap clear of the water when swimming. Flying fish can glide for several hundred yards over the surface of the sea when escaping from predators. Salmon can even jump over waterfalls when swimming up rivers to breed.

Some fish can move about completely out of water. Eels sometimes do a "cross country" run from one pond to another. Mudskippers crawl over the exposed mud of a mangrove swamp at low tide searching for food.

▲ Although beautifully streamlined for fast swimming, a shark has problems staying afloat. If a shark stops swimming, it sinks. This is because sharks do not have a swim bladder to make them buoyant. They stay afloat by using their front fins like hydrofoils. Some sharks actually store large amounts of oil in their huge livers. You can probably guess why.

▶ A seahorse does not move by waggling its tail like most fish. It gently waves its dorsal fin and this pushes the seahorse forwards in an upright position.

It takes all shapes

A streamlined shape is ideal for fast forward movement in a straight line. But some fish need to maneuver or twist and turn quickly. They need a different shape and rely less on power from the tail.

Fish that live on the sea bed, like skates and rays, have a flat shape. They swim along very slowly, and rely mainly on their pectoral fins to help move them along.

▶ This huge manta ray swims slowly by flapping its huge pectoral fins up and down. It seems to "fly" underwater like a giant bird.

More about ▶▶ Streamlined shape p 20-21, 36 Muscles and bones p 8-9, 37 Mudskippers p 32 Sharks p 17

Flippers and flukes

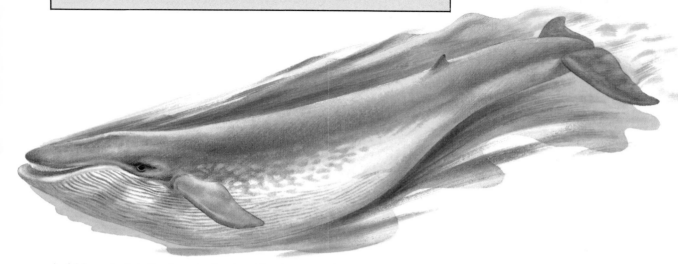

▲ A blue whale is the heaviest animal that has ever lived. A fully grown female is over 90 feet (30 meters) long and weighs about 150 tons (150 tonnes). This is about 30 times heavier than an African bull elephant, the world's biggest land animal.

You would expect to find the biggest fish living in water. But does it surprise you that the heaviest amphibian, reptile, and mammal are also aquatic, or water-living, animals?

As water is much denser than air, it holds things up or supports them better. An animal like the blue whale can only live in water. It is too big and heavy to support itself and move on land.

Shaping up

Whales and their relatives, porpoises and dolphins, are wrapped in a layer of blubber, or thick fat. It keeps them warm and helps them float more easily. It also fills in all the hollows and evens out all the bumps on their skin so they can slip smoothly through the water.

▼ Swimming movements in whales are powered by large muscles lying above and below the backbone. They contract, or shorten, in turn and move the tail up and down. It is a kind of up and down skulling movement that moves the whale forward.

▲ A whale's tail fin or fluke is arranged horizontally, and not vertically as in a fish.

Jumping out of your skin

It took scientists a long time to understand why dolphins are able to swim so quickly. They can move much faster than expected, even though they are beautifully streamlined. It has now been discovered that they accelerate by "jumping out of their skins." As they move forward, they leave behind a very thin skin "ghost" of themselves. The drag of the water works on this as the dolphin escapes. It's like a slippery piece of soap popping out of your wet hand.

Clumsy "landlubbers"

Seals move very awkwardly on land. Sea lions can lift themselves up on their flippers and shuffle on "all fours." They can even gallop by moving front and back flippers alternately. But true seals are much more clumsy. They crawl on their bellies, humping themselves along. Walruses do both. They shuffle on their front flippers and help themselves along with their big tusks.

Seals, sea lions and walruses

Seals and walruses are also streamlined with arms and legs formed into flippers. Sea lions "fly" underwater like penguins by moving their front flippers up and down. The back flippers are used mainly for steering.

True seals like the grey seal use only their back flippers when swimming. They also swing the body from side to side as they move along. Walruses also use their back flippers to push their huge bodies through the water.

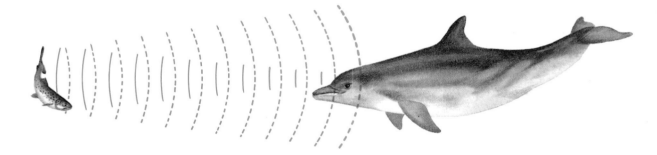

▲ Dolphins use a sonar system like bats to find their way about. They make high-pitched clicks. A dolphin makes about one click every second when swimming gently. The clicks speed up to about 500 a second when it chases its prey.

▶ Seals have an exceptionally streamlined shape. They have tiny ear flaps, or none at all. The male's reproductive parts lie inside and even the female's nipples are flush with the body. Seals leave nothing sticking out to spoil their extra smooth outline.

More about ⟫⟫ Streamlined shape p 18-19, 36 Muscles and bones p 8-9, 16-17, 37 Sonar systems p 55

Walking on water

The halfway world between air and water is a strange place to live and move in. Animals that live there have found many different ways to travel about. Jellyfish such as the Portuguese man-of-war allow themselves to be blown by the wind. Others, like flying fish, travel just above the surface for short periods before flopping back into the sea. Many water birds paddle around on the surface for much of their lives. There are even some animals that have become specialized for living and moving on the surface of the water. They move at the meeting place of air and water.

Running on water

Some animals move by water skiing or skittering. Frogs sometimes skip or bounce across the water surface in a series of quick jumps.

A basilisk lizard from South America has extra strips of skin on the sides of its toes. It uses its large, flat feet to run across the surface of ponds and small streams. Can you guess why it is sometimes called the 'Jesus Christ lizard'?

▶ Skippers skitter across the surface of the sea. They beat their tails in the water and travel upright with the rest of their body in air. Skippers probably use this way of moving to escape from predators.

Lily trotter

No bird can actually walk on water but the African jacana does the next best thing. Its big feet spread its weight out over a wide area. This allows the bird to walk across the floating leaves of plants without it sinking. No prizes for guessing why the jacana's other name is lily trotter!

◀ The African jacana looks a very clumsy bird. Its huge feet seem out of proportion to the rest of its body.

The world of film

Water has a kind of thin elastic skin on its surface like the skin that forms on a bowl of cold custard. This film has a pulling power called surface tension. Some animals live and move on this springy, bouncy platform or surface film. Any animal moving on this thin skin must be small and light enough not to break it.

Pond skaters, whirligig beetles and spiders all live in this strange halfway world between water and air. They have special adaptations such as wax-tipped feet. Wax and water do not mix. They actually repel or push each other away. Waxy feet help these small animals not to break the surface film.

▲ A pond skater walks on water by spreading out its long thin legs. They spread its weight over a big area. You can see that the pond skater's feet do not break the surface film. But each foot makes a small dimple on the film as it pushes on to it.

Champion water skier

The camphor beetle "skis" across the surface film of water at great speed. It produces a liquid from its abdomen or back end. It is a special liquid which lowers the surface tension of the water behind it. The stronger pull of the film in front now drags the beetle forward. It zooms along as if its body is driven by a tiny outboard motor. The beetle can even steer itself by waggling its abdomen from side to side.

Spring to it!

Springtails are tiny insects less than $\frac{1}{8}$ in (2 mm) long. They are some of the most unusual movers in the animal world. Underneath the body there is a tiny spring-like structure held by a catch or trigger. It is a built-in "pogo stick." When it wants to move, the springtail undoes the tiny catch and releases its springy pogo stick. Its body shoots up into the air, to many times its own height. Springtails can often be seen bouncing up and down on the surface film of ponds.

More about ⟩⟩ Flying fish p 49 Frogs p 26-27
Jacana p 40 Beetles p 51

Creepy crawlies

The animals we are most familiar with walk on four legs. But do not be misled. Most animals on the Earth's surface are insects and these walk on six legs. Some animals such as centipedes and millipedes walk on hundreds of legs. So walking on four legs is really quite unusual!

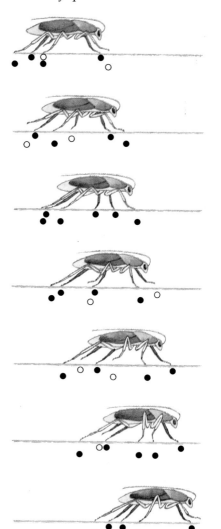

Life on six legs

Insects have multi-jointed tube-shaped legs that are moved by muscles inside. The six-legged design makes insects very stable animals. You rarely see an insect fall over! When an insect walks, its front legs reach forward while its back legs push against the ground. The middle legs also remain on the ground and work like small pivots to help the back legs lever the body forward.

◀ These drawings show the walking cycle of an insect like a cockroach. The black dots show which of the insect's legs are on the ground. The white dots represent the legs in the air.

▲ Each leg of an insect ends in a pair of tiny hooks or claws with a pad between them. These help the insect to walk on any kind of surface, rough or smooth. A fly has no difficulty in walking upside down on the ceiling.

▲ Spiders move by a combination of muscles and fluid power. The muscles bend the legs and then the spider pumps fluid in to straighten them again.

▶ A millipede's legs move in tiny wave patterns. Small groups of legs are moving forwards while others are swinging backwards. But there are always some legs in contact with the ground.

Lots of legs

Why don't animals such as centipedes and millipedes, which have lots of legs, trip over their own feet? On some occasions, we have difficulty with just two! The answer is that the movements of the four hundred or so legs of a millipede are very carefully controlled by a complicated nervous system.

Even though a millipede has lots of legs, it does not move very quickly. This is because its legs are short and its body is close to the ground. Each leg has only a short stride. When a millipede walks, each foot touches the ground just before the one in front. The legs move like the keys of a piano when you draw your finger quickly across them.

A soft option

Not all "creepy crawlies" have legs. Even though soft-bodied earthworms are legless, they have no difficulty burrowing through the soil.

Another very large group of soft-bodied animals, the mollusks, have developed lots of ingenious ways of getting around. Octopuses and squids power themselves by a kind of jet propulsion. But instead of using air, they make do with a jet of water.

Snails and their relatives move on top of a big piece of muscle called a foot which is shaped rather like your own tongue. Muscular ripples pass down the foot from front to back and carry the snail along on a kind of conveyor belt.

▼ Some mollusks use their foot to flick themselves forwards.

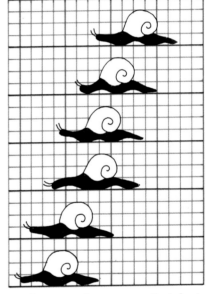

▲ Some snails even use their foot to lever themselves along. They "gallop" at a surprising speed.

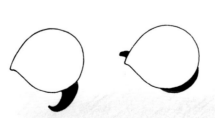

Hopping and jumping

If you have ever tried hopping and jumping, you will know that they are both very tiring exercises. Next time you go out, try hopping down the street on one leg. See how far you travel before you become exhausted. Your journey will be even shorter if you try hopping on both legs together!

Many animals move mainly by hopping and jumping. It is easy to tell a "hopper" or a "jumper" by looking at the animals' legs. Good jumpers have powerful back legs to push themselves quickly upward and forward.

Jump to it

Frogs are expert jumpers with back legs longer than the body. They usually jump to escape from predators, but tree frogs also sometimes leap and snap at flying insects. When resting, a frog's back legs lie folded under its body.

When it wants to make a sudden escape, a frog quickly stretches its back legs and pushes against the ground.

▲ The kangaroo can jump further than any other animal. It has enormous feet, up to 16 in (40 cm) long. It uses its feet like two spring-boards to bound along at speeds of 25 miles an hour (40 kilometers an hour). A kangaroo's legs are so powerful that it can cover 33 feet (10 meters) in a single jump. At high speed, its long tail streams out behind as a counterbalance.

This makes it move upward and forward. A frog often pulls its eyes into its mouth when it jumps. This protects its eyes and also helps to make the frog more streamlined.

▼ When a frog jumps, it follows a curved path through the air. Although it takes off on its back legs, a frog always lands front feet first. The front legs are shorter than the back ones and act as shock absorbers to help the frog make a safe landing.

Nature's oddity

The strange kangaroo rat looks like one of nature's design mistakes. But its outsize back legs, enormous feet, and long bushy tail makes the kangaroo rat a perfect hopper and jumper. It can move very quickly and is easily able to cover 16 feet (5 meters) in less than a second.

Long ankles are the trade mark of the Malaysian tarsier. These little tree-dwelling mammals can jump 5 feet (1½ meters) high and they can leap more than 7 feet (2 meters) in a single bound. Just before it lands, a tarsier pulls its long bushy tail over its back like a parachute to slow it down for a safe landing.

▲ A kangaroo rat's tail is three times longer than its body. Using its tail like a rudder, a kangaroo rat can change direction in mid hop. It can even make a 90-degree turn while leaping through the air.

Flick-a-pic

1 Trace the outline of each frog on to a separate piece of paper.

2 Stick each drawing on to a piece of thin card. Each piece of card should be the same size.

3 Staple the pieces of card together, and you have a small flick book.

Can you make your frog come to life by flicking the pages of your book?

A grasshopper has big back legs for jumping. They are worked by powerful muscles which straighten the legs to lift the animal high above the ground. Grasshoppers leap to escape danger. But they also leap to get from one place to another. It is quicker than walking!

A flea uses its back legs to jump more than 200 times its own length. Little elastic pads at the base of its legs catapult it into the air.

More about >> Frogs p 22, 33, 49 Grasshoppers p 8

Snakes on the move

Snakes at Sea

Many snakes are excellent swimmers and some spend their lives in water. When a snake moves in water it swims like an eel. It "wriggles" by bending its body from side to side. As the body waves pass from head to tail, they push against the water and this drives the snake forward.

◀ Pythons and boas still have the remains of their back legs but they are hard to see.

Snakes live in many kinds of habitats including burrows underground, on the ground itself and in trees. There are even some snakes that spend their lives at sea! The heaviest of all snakes, the anaconda, also stays in water for long periods and is an excellent swimmer. All snakes have a long, cylindrical body with no legs. As snakes have no legs, they have developed other ways of moving around.

Body bending

A snake has large muscles running down its body on each side of its backbone. When a snake moves, the muscles on one side become shorter while those on the other side are stretched. By using its muscles in this way, a snake turns its body into a series of waves or bends. The underneath of a snake's body is covered in thick scales. As the waves pass down the body, the scales push against the rough surface on which the snake is lying. This pushes the snake forward.

▼ A snake glides along in a series of S-shaped bends. But it can move only on a rough surface. If a snake tried to move on a completely smooth surface, it would wriggle around helplessly and get nowhere.

Going straight

Big snakes like boas, pythons and vipers have developed a special way of moving. These snakes have a powerful band of muscle running underneath the body from head to tail. The large belly scales are fixed to this ribbon of muscle. When the muscle shortens, the snake glides forward on its scales. It works like a caterpillar track, carrying the snake on top. Straight-line movement like this is used by large, heavy snakes when they are not in a hurry. A boa or python speeds up by moving in the usual way.

The concertina method

Some snakes use another method for climbing trees with a smooth trunk. A tree boa coils its body around the base of the tree. It then holds on with its tail and reaches up with its head to hook its neck higher up the trunk. The snake then lets go with its tail and pulls its back end up to where its neck is holding on. By repeating these steps, the snake can climb even the smoothest tree trunk.

▶ Sidewinder snakes move at 2–2.5 miles an hour (3–4 kilometers an hour). As most of the snake's body is held off the surface of the hot sand, this way of moving also helps to keep the body cool.

A sideways move

The sidewinder rattlesnake uses a remarkable technique for moving over sandy ground. Its body moves in an S-shape but it touches the sand only in two places, near the head and tail. These two points travel very quickly down the body. The snake keeps throwing these loops both forward and sideways.

As it moves, the snake leaves behind a record of where its body has been. Its tracks look like a series of bars in the sand. The marks lie at an angle of 45 degrees to the direction in which the snake has actually traveled.

Weight and balance

Before an animal walks or runs, it has to learn to balance. An animal with four legs is a bit like a table. Look at the pictures. The table on the left does not fall over because it has four legs, one at each corner. The table's weight is spread evenly over each of the four legs.

If we take away one of the legs, the table falls over. This is because its weight is no longer spread evenly. But we can make a three-legged table stand up. All we need to do is move the heavy weight nearer to the corner opposite from the missing leg.

Tables and animals

Now let's swop the table for a four-legged animal. When our animal stands up, its weight is spread evenly over its four legs. The animal is balanced. But what happens if the animal wants to lift one of its feet off the ground? Does it fall over? A three-legged table can be made to stand up if its weight is moved. What do you think an animal must do before it lifts one foot up?

Did you know that every object has a center of gravity?

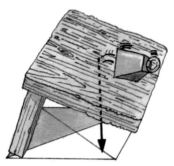
Yes, the center of gravity is the point around which an object's weight is evenly spread

This table's center of gravity falls inside the shaded area. This is why the table stands upright.

This table's center of gravity falls outside the shaded area. This is why the table falls over.

Now the center of gravity of each table falls inside the shaded area. This is why the tables stand up.

BOTH TABLES STAND UP, EVEN THOUGH EACH HAS ONLY THREE LEGS.

Where is this animal's center of gravity?

A   B

▲ Animal **A** shifts its weight towards its tail end before lifting one of its legs. Animal **B** shifts its weight towards its head end before lifting one of its back legs.

Front and back

Some animals have their center of gravity nearer the front legs. Others have it nearer the back legs. A horse belongs to the first group. This is why a horse finds it easy to kick with its back legs. Bears, rabbits, and squirrels belong to the other main group. This is why a grizzly bear can easily rear up on its back legs, and why hares stand up to "box" each other.

Throwing your weight around

Animals lift their feet off the ground when they walk but they seldom fall over. They balance themselves by moving their weight. You can see how they do this by looking at the two drawings of a four-footed animal.

You can do a similar thing. But in your case it is a bit more difficult because you have only two legs. When you lift your left foot, you automatically transfer your weight to your right foot and vice versa.

The slip before a fall

Sometimes you make such a sudden movement with one of your legs that you do not have time to move your weight in order to balance yourself. When this happens, you slip and fall over.

▲ This baby giraffe has not yet learnt to balance. But it quickly learns how to transfer its weight from one foot to another. It will be on its feet and running soon after birth.

More about ⟫ Walking p 34-35 Horse movement p 33, 35
Giraffe movement p 35

Lift off

Animals living in water are not affected very much by gravity. The water supports their weight. But for land animals it is a different story. An animal moving on land has to support its own weight, as well as transport it about.

In days of old

The first animals began to leave the water about 200 million years ago. These land invaders were small fish-like animals that wriggled about in prehistoric swamps.

These animals had no legs to walk on. Instead, they dragged themselves along on their front fins. There are some fish alive today which still move in this old-fashioned way, just like the earliest land walkers millions of years ago.

▲ Mudskippers live in tropical mangrove swamps. At low tide they wriggle about on the surface of the mud looking for food. A mudskipper props itself up on its front fins, which it uses like a pair of small crutches to jerk itself along.

Up and under

Dragging the body along is a very inefficient way of getting about. It is slow and it uses up a lot of energy. Moving becomes much easier and quicker if the animal's body is lifted clear of the ground. Early amphibians and reptiles were the first to try this and today's toads, newts, and lizards still move in a similar way. They walk in a kind of double-ended push-up position.

◀ A crocodile still walks and runs like its early ancestors millions of years ago. It lifts itself up on all fours so that its body is clear of the ground. In this position it can move quite quickly. But it soon becomes tired and flops down to rest.

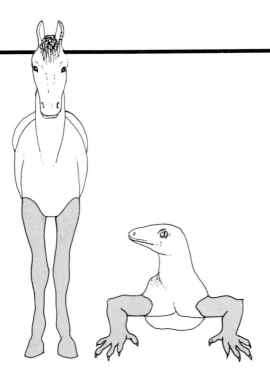

To sag or not to sag

A horse's weight lies directly over its legs when it stands on all fours. This arrangement gives good support against the force of gravity. This is why a horse can stand up for long periods without getting tired. The legs of a lizard are arranged differently. A lizard's weight is not carried directly above its legs. Its body sags between them. A lizard's legs give much poorer support. You will know how a lizard feels if you have ever tried to do some push ups on the floor!

▲ When it comes to standing, gravity affects a horse much less than a lizard. This is because of the way its legs are positioned.

▼ This maned-wolf from South America has extra long legs. Long legs like this would be no good stuck on the sides of its body. But positioned underneath they allow the wolf to stand up and run for long periods without getting tired.

Walk before you run

You cannot run before you can walk, and you cannot walk before you can stand. Standing is the first really important step to moving efficiently on land. Reptiles and amphibians have their legs on the side of their bodies. They usually rest with the body flat on the ground. They stand up on their legs when they want to move. But amphibians and reptiles use a lot of energy standing. This is because gravity keeps pulling them down. This is why frogs and lizards cannot stand up for long periods and why they soon become tired after moving about for a short time.

More about ⟩⟩ Support p 20 Mudskippers p 19
Frogs p 22, 26-27, 49 Horse p 31, 34-35

Walking and running

Animals are designed or adapted for walking and running in different habitats. You can usually tell what kind of surface an animal moves on by looking at its legs. Their length and their shape are very important.

▶ Animals like these zebras have quite short bodies and long legs. They are built for running on flat surfaces. How would you describe a squirrel's body and legs?

The horse and the squirrel

1 Roll a cylinder out of a piece of plastic clay. Make your cylinder 1.6 in (4 cm) long and 0.8 in (2 cm) in diameter.

2 Cut a drinking straw into four pieces, each 1.6 in (4 cm) long, and stick each piece into the plasticine cylinder. You now have a model horse with four legs. Test your model to make sure it stands up properly.

3 You can make a cardboard head and tail to make your model look more lifelike.

4 Stand your model horse on a book or other flat surface and gradually tilt it so that your model is standing on a slope.

How far can you tilt your model horse before it falls over. Measure the angle of the slope with a protractor.

5 Now repeat 1 to 4 with a model squirrel. Make its body 2.4 in (6 cm) by 0.4 in (1 cm) and each leg 0.4 in (1 cm) long. Can your squirrel stand on a steeper slope than your horse?

Hint
Your experiment will work better if you stick a blob of plastic clay on the surface behind each back leg. This will stop your model from slipping backwards.

Left right, left right

When any four-footed animal walks slowly and steadily, it lifts only one foot off the ground at a time. This is true of a toad, a chameleon or an elephant, or even a young baby on all fours.

If you watch an animal walk, you can see that it moves its legs in a particular order or sequence. So, starting with the left foot, it moves left front, right back, right front, left back. Then it repeats the sequence. Try crawling slowly on all fours and you will see how this works.

Look, no feet!

Most animals move faster than a slow walking pace. When they increase their speed, the pattern of leg movements begins to change. Look at the pictures which show a horse walking.

When a horse gallops, its leg movements change again. You can see a horse galloping in the pictures below. You will find it difficult to work out the sequence of leg movements. But you can see that sometimes it has only one foot on the ground and sometimes it has none at all!

▼ A horse never walks really slowly. Therefore it never goes left, right, left, right. The picture below shows a horse walking. Can you work out the order in which the horse lifts its feet off the ground. It is a bit more complicated than a slow walk.

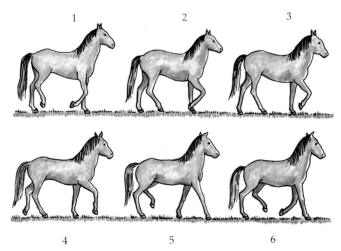

On your toes

Animals like bears walk with their big feet flat on the ground. We do the same. First the heel goes down and then the ball of the foot and the toes. This is fine for walking, but it is not much use for fast running. Think what you do when you sprint to catch a bus. You run on your toes. Fast-running animals like horses and antelopes walk and run on their toes. Their long legs also help them gain extra speed.

▲ When you visit a zoo watch a camel or giraffe walk. Do they move their legs in the usual way?

More about ⟩⟩ Horse p 31, 33 Giraffe p 31
Human walking p 38-39

Super sprinter

The world's fastest

The cheetah is the fastest animal on land. Its method of hunting is very simple. It depends on sheer speed to catch its prey. A cheetah can run at speeds close to 60 miles an hour (100 kilometers an hour). But, like all good sprinters, it is unable to keep up such a fast speed over very long distances. During a chase, a cheetah sprints flat out only for the last 330 to 440 yards (300 to 400 meters) when it is closing in on its kill. If a cheetah misses its prey, it gives up and has a rest before trying again.

Built for speed

A cheetah's body is perfectly built for high-speed running. It is light in weight. The average cheetah weighs less than 132 lb (60 kg). Its legs are long and thin with powerful muscles to move them.

A cheetah cannot pull its claws in completely like your pet cat. When running at full speed, its claws stick out to give extra grip on the ground. This prevents it from slipping or sliding. Even a cheetah's ears are smaller than usual. This also helps to reduce air resistance at high speed.

▲ This is a cheetah running at full speed. You can see all four legs are off the ground. The animal's body has also become streamlined to help increase its speed by reducing air resistance.

Why so fast?

The cheetah's backbone holds the secret to its amazing speed. It is a very flexible structure which bends up and down as the cheetah runs. This allows the cheetah to take extra long strides.

The cheetah's shoulder blades and hip bones are also flexible. They can swivel backwards and forwards easily to help the legs stretch out even further. Because the backbone, shoulder, and hip bones move in this way, the cheetah's body is very supple and loose. A relaxed body helps the cheetah to run more quickly.

The tail end

A cheetah's tail is longer than its body. It helps balance the cheetah and works like a rudder. It makes the animal's body more stable during a high-speed sprint. It also keeps the cheetah on course. Running at 60 miles an hour (100 kilometers an hour) can be dangerous. If a cheetah lost control and fell over, it could seriously damage itself. Even the tail's white tip is important. Baby cheetahs use it as a signal to help them keep up with their mothers.

▲ When a cheetah is bunched up, its backbone curves upwards. The cheetah's body is like a tight spring ready to uncoil. As it leaps forward, the "spring" is released. Now the backbone curves downwards so that the cheetah can take the longest stride possible.

▼ Between each leap and bound, a cheetah's body is completely off the ground twice. This happens when it is fully bunched and then fully stretched. The distance covered before the same foot touches the ground again can be more than 23 feet (7 meters).

More about ⟫ Streamlining p 18-19, 20-21 Backbones p 18, 20
Tails p 44-45

Life on two legs

Balancing and walking on two legs is much more difficult than standing and moving on four legs. Even so, some animals do occasionally move on two legs. Both humans and birds are completely two-legged when it comes to moving on land.

In some animals the back legs are bigger and more powerful than the front pair. These animals are part-time two-legged movers. They are hoppers, leapers, and jumpers. So, although animals like kangaroos and frogs can walk on four legs, when they want to make a fast getaway, they become two-legged hoppers.

▲ The ostrich is the fastest two-legged runner in the animal world and can reach speeds of nearly 45 miles an hour (70 kilometers an hour). Each foot has only two toes and one is much bigger than the other. An ostrich really runs on the one big toe on each foot!

▲ The Australian bearded-lizard is a bit of a puzzle. It moves faster on all fours, but when the lizard wants to escape from predators it dashes off on its back legs. Running is hot work. Perhaps the lizard keeps cooler by turning into a two-legged sprinter.

A question of balance

A two-legged animal needs to balance even more skillfully than one with four legs. The human backbone is specially curved to bring the weight of the body over the legs and feet. This makes balancing on two legs much easier.

One for the road

The roadrunner lives in desert regions in North America. It runs very fast and can easily reach a speed of 19 miles an hour (30 kilometers an hour). Its long tail helps it to balance at high speed. Its tail also works in the same way as a steering wheel. A quick flick of its tail to one side or the other steers the roadrunner onto a new course.

The human story

All apes can walk on their back legs for short periods, but they usually end up shuffling along on all fours. But humans are really two-legged walkers or bipedal. A young baby first starts to move by crawling on all fours. Then, as the child gets older, it begins to practice balancing and walking on two legs. We have to learn to move our weight from one foot to the other, just like a four-footed animal.

▲ The roadrunner is a long-distance sprint champion. With the help of its rudder-like tail it can zig-zag across country at high speed. When it wants to brake, it flicks its tail over its back and comes to a sudden stop.

The spotted skunk has an unusual way of threatening intruders. It does a hand stand and then sprays out an awful-smelling liquid from a gland on its bottom.

◄ Gibbons balance with their arms when walking along forest vines. This drawing is based on a series of photographs taken in a rain forest in south-east Asia.

More about 〉〉 Hopping and jumping p 26-27 Kangaroos p 26
Frogs p 26-27 Human walking p 34-35 Gibbons p 42-43

Softly softly

Have you ever tried walking or running on a sandy beach? If so, you will know how difficult it is. You keep sinking into the soft sand. You also get tired very quickly. You probably try to get on to firmer ground as soon as you can. It is the same walking in deep snow and thick mud.

Animals that live in deserts, swamps or where there is plenty of snow often have specially designed feet. They are flat with widely spread toes. They are designed to spread the animal's weight over a wide area and to stop it from sinking or getting stuck.

▲ The sitatunga is an antelope from Africa. It lives in swamps where the ground is very soft. When it walks, it spreads its hooves wide apart to prevent it from sinking into the soft, boggy ground.

◀ The flamingo has webbed feet to help it swim. The webs between its toes give the feet a large surface to push the flamingo through the water. The large flat feet also help the flamingo to walk over soft mud without getting bogged down.

Stick in the mud!

Many water birds have to walk over the soft ground in swamps and on river banks and lake edges. Many of them have feet adapted for going "softly softly." Coots have fringed toes to give them a bigger surface for mud-walking. The jacana's extra long, thin toes look very comical. But they are the perfect adaptation for walking on floating lily pads. Mammals like otters use their webbed feet like "mud shoes" to prevent them from becoming a "stick in the mud."

▲ The clown dune cricket lives in the Namib Desert in Africa. It has "flower-like" feet to help it move across the sand.

Shifting sands

Hot, sandy deserts are not easy places to live and move in. Animals that are found there often have specially adapted feet for walking on the soft ground underfoot.

A camel's widely-splayed feet allow it to walk in even the deepest sand. The little jumping jerboas have hairy back feet for the same reason. The addax antelope has big hooves, like its relative the sitatunga. This time, however, they are for walking on sand, not mud!

Snow shoes

Walking on snow is just as difficult as moving over soft mud. Animals need feet designed specially for the job. The ptarmigan is a bird like a grouse which lives in cold parts of the world. It has fluffy, feathery feet which support it on soft snow.

The snow leopard spends all its time in mountainous areas where there is plenty of snow. Its large, padded feet are covered with thick hair. They work like two pairs of snow shoes when the leopard walks in deep snow.

There is even an animal called the snow shoe hare. Can you guess what its feet look like?

▲ A reindeer has feet like large, flat plates. They spread the animal's weight over a wide area and stop it from becoming snow bound.

▲ Each foot of a camel is made of two big toes. The huge, padded feet distribute the camel's weight so that walking on soft sand is easy.

Animals sometimes find other ways of moving on snow and sand. When adélie penguins get tired of walking, they slide down snowy slopes like little toboggans. Rather than walk, some desert lizards "swim" through their sandy home by wriggling their tails like fish.

More about ⟩⟩ Webbed feet p 16 Camel movement p 35

King of swing

Many different animals live high up in the leafy canopy of a tropical rainforest. Some move around slowly by clambering, climbing or crawling. Others get around more quickly by hopping, jumping or flying. But, apart from birds, no animals move so quickly, or as gracefully, as the gibbons in the leaf canopy.

These small apes, which come from south-east Asia, have developed a method of locomotion called brachiation. They swing hand over hand from branch to branch on extra long, strong arms. A gibbon can move through tree tops almost as quickly as a human can run on the ground below.

▲ Gibbons launch themselves into the air as they leap from one tree to another. They make "rowing" movements with their legs as they speed between the branches, rather like a long jumper doing a series of hitch kicks. These leg movements help to propel the gibbon on to its next handhold.

Make or break

A gibbon's hands are like a pair of flexible hooks. As the gibbon "flies" through the trees its hands can quickly grasp any branch. Gibbons are so agile that they even catch birds in mid-air! But although they are remarkably skilled acrobats, they often have accidents when they miss a handhold and fall. Scientists studying gibbon skeletons in museums often find broken limb bones.

▼ A gibbon swings hand over hand as it moves along at great speed.

Old man of the forest

Orang-utans are also skilled acrobats. They live in the forests of Borneo and Sumatra where they spend their life climbing among the trees.

Orang-utans are much bigger and heavier than gibbons. An old male orang-utan may weigh as much as 22 lbs (100 kg). You need extra strong arms to lift such a heavy weight as this. With its arm span of more than 8.25 feet (2.5 meters), an orang-utan can cover large distances in the tree tops very quickly.

► Apart from brachiating, orang-utans are also expert climbers. They often use their feet to get an extra toe-hold.

Flick-a-pic

To see how to make this gibbon come to life see p.27

More about ⟫ Gibbons p 39 Clinging and climbing p 44-47

Tail tales

Moving about in trees high above the ground can be a dangerous business. It helps to have a bit of extra help! Many animals that live in trees have a long tail which they use like another arm or leg. It forms a kind of fifth limb which they use for holding on with and for gripping.

Tails you hold

We call a tail used for holding on with a prehensile tail. South American monkeys like spiders monkeys and woolly monkeys climb quickly through the jungle canopy using their tails for extra grip. They can even hang by their tail while they feed.

Tail-tale prints

Every spider monkey has a bare patch of skin underneath its tail at the tip. The surface is covered in tiny ridges, just like the skin on your finger tips. This rough surface helps the monkey to get a better grip with its tail. Each spider monkey has its own special pattern of ridges. They are the monkey's "tail prints."

▶ The spider monkey is the most agile animal in the Brazilian jungle. Its tail is one and a half times longer than its body and thicker than any of its limbs. It is a very strong tail and can easily support the monkey's weight as it moves about.

A spider monkey's tail is so sensitive it can pick up something as small as a peanut. Their tails are so important to spider monkeys that they even hold tails instead of hands. Babies often hold tails with their mothers.

▶ This chameleon looks as if it has three back legs. It is using its tail to hold on with as it tries to catch an insect with its long tongue.

Snakes and lizards

Nearly all reptiles that live in trees have prehensile tails. Tree snakes grip with their tails and then shoot out the rest of their body when they strike at another animal.

Some types of gecko lizards have scales or ridges on the underside of their prehensile tails, rather like spider monkeys. They work like tiny suction pads, helping the geckos to get a better grip with their tails.

Some lizards have a single scale shaped like a small claw at the tip of their tail. This probably helps them to get a better grip with their tails when climbing.

▲ The tree porcupine lives in the jungles of South America. Its prehensile tail is different from that of other animals. It curls upwards rather than downwards.

▲ The emerald tree boa coils its tail and most of its body around the branch on which it is resting.

▲ A seahorse is the only fish with a prehensile tail. But it does not use it for moving. Instead, the tail acts like an anchor. It prevents the seahorse from being washed away by underwater currents.

More about ⟫ Clinging and climbing p 42-43, 46-47 Snakes p 28-29 Seahorse p 19

Clinging and climbing

Many kinds of animal spend their lives moving, feeding and sleeping above the ground. Some animals climb among the vegetation. Others cling and climb on the vertical sides of rocks and tree trunks. Some animals like houseflies and geckos can even move upside down on horizontal surfaces. All these animals have feet specially adapted for clinging and climbing.

Non-slip feet
Mountain goats can leap from rock to rock and climb very steep slopes. Their hoofs have very sharp edges which catch in the crevices in the rock. Their hollow soles also act like suction pads.

▲ A gecko's feet give it more than 100 million points of contact with the surface on which it is clinging. The hold may be so strong that any attempt to pull a gecko off a sheet of glass may result in the glass breaking before the gecko lets go.

◄ Tree frogs have large suction pads on each finger and toe. These pads are moved by special muscles. The surface of the pads produces a glue-like substance which helps the frog to remain lightly stuck to the underneath of leaves for long periods.

Super glue
Geckos are remarkable climbers. They can easily walk on a vertical surface and can even walk upside down across the ceiling. Each of a gecko's toes has a flat pad covered in small ridge-like scales. The scales have thousands of tiny hair-like brushes on their surface. These brushes push into all the little nooks and crannies present even on the smoothest surface. This is how a gecko gets its super grip.

As slow as a sloth

Sloths live in the rain forests of Central and South America. They are the slowest of all mammals. A mother sloth was timed moving at 13 feet an hour (4 meters an hour) when hurrying towards her baby!

A sloth never moves more than one limb at a time. It has long, thin hands and feet and its fingers are bound together by a kind of elastic material. It can hardly move its sharp claws, so they make pefect hooks for clinging on with. A sloth hangs upside down, even when it sleeps.

Three and one or two and two

The foot of a perching bird has three forward-facing toes and one that points backwards. Feet like this are ideal for gripping, even when the bird is asleep.

Some birds, like parrots and woodpeckers, can climb up and down tree trunks. Their feet have a different shape compared to those of ordinary birds. Each foot has two toes in front and two behind. Parrots also use their beak as a third "foot" to help them climb.

Squirrels and some lizards can swivel their back feet to face the other way. This helps them to get a better grip when they are climbing down very steep surfaces.

▶ Woodpeckers use their extra stiff tail feathers to help prop themselves against a tree trunk on which they are climbing. They act like an anchor to stop the bird from sliding back down.

Power grip

Animals that live high above the ground often have hands and feet designed for grasping or holding. Apes and monkeys are able to move quickly because their hands and feet grasp the branches with a firm grip. Their smaller relatives, the pottos and lorises, have short, stubby fingers. These allow them to grip even more powerfully. A slow loris has no difficulty in hanging upside down by one leg!

More about ⟫ Squirrels p 34 Gripping p 42-45 Frogs p 26-27

Living parachutes

Defying gravity is a dangerous thing to do. To move safely through air one requires a specially designed body. If an animal wants to remain airborne it needs wings or flaps of skin to give it lift. It also needs to be made of lightweight materials.

Insects, birds, and bats are good fliers. They have all developed powered flight in which the wings are moved by strong muscles. If an animal cannot flap its wings, the only other way to fly is to glide. There are some very good gliders in the animal world.

Membranes

Gliding animals include fish, amphibians, birds, reptiles, and mammals. There is even a gliding snake! In order to glide properly, you need to fall gently from a high point to a lower one. In order to slow down their falling speed, gliders have developed webs or flaps of skin to increase their body surface. These flaps of skin work like parachutes once the animals have launched themselves into space.

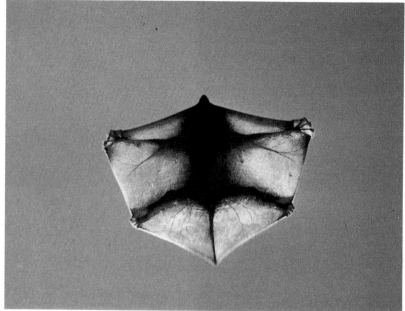

▲ The colugo looks like a kite as it glides through the air. It has the biggest membrane of any gliding animal. A colugo can glide as much as 148 yards (135 meters) between trees and lose very little height. When a female colugo is resting, it folds part of its membrane into a soft warm pouch. A baby colugo can rest safely in its mother's hammock!

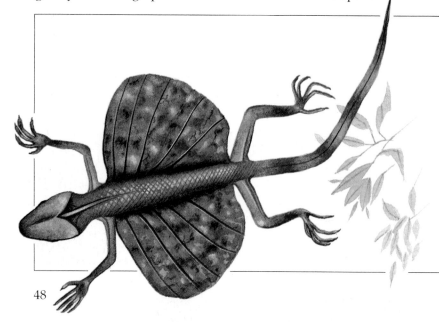

The draco lizard lives in the forests of south-east Asia. Its "wings" are made of thin membranes which are strengthened by five ribs on each side. Although this small lizard cannot glide as far as a colugo, it can make complicated mid-air movements. When resting on a tree trunk, the "wings" are folded away.

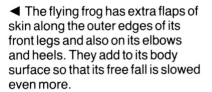

◄ The flying frog has extra flaps of skin along the outer edges of its front legs and also on its elbows and heels. They add to its body surface so that its free fall is slowed even more.

The free-fall frog

A very strange frog lives in the canopy of the jungles of Borneo and Sumatra. It is a small tree frog with thin legs and enormous, webbed feet.

It is a gliding frog and it uses its feet like four parachutes to slow its fall as it leaps from the tree tops. By spreading the webs between its toes, it doubles its body surface. Flying frogs can glide up to 37 feet (12 meters) before landing on a tree lower down. They can even steer and change direction by moving their legs and by altering the shape of their webbed feet.

Gliding snakes

Visitors to the rainforests of south-east Asia sometimes see one of the most remarkable sights in the animal world — a flying snake!

The paradise tree snake can glide more than 44 yards (40 meters) through its tree-top home. As it lauches itself into space, the snake sails through the air in an S-shape, using its tail as a rudder. It hollows its body underneath to trap a layer of air which cushions its fall. The paradise snake "flies" to escape from its predators and to chase its own prey from tree to tree.

On gossamer wings

When young spiders are ready to live on their own, they use a method of ballooning to find new homes. They spin silk strands called gossamers which catch the wind and float up into the air, carrying the small spiders with them. Adult spiders also sometimes travel by "ballooning" to reach new places. Sometimes they are carried to remote places thousands of kilometers away.

Flying tonight

Flying fish escape from predators by leaving them behind in the sea. They taxi along on the surface by wagging their tails in the water, their huge front fins held out like a pair of wings. When they reach a speed of about 37 miles an hour (60 kilometers an hour), they take off and glide for about 550 yards (500 meters) before flopping back into the water.

◄ These flying fish are probably trying to escape from predators below the surface. Dolphin fish often follow the flight path of flying fish and catch them as they land back in the water.

More about ⟫ Flying p 50-55 Gliding p 53 Frogs p 26-27
Snakes p 28-29 Spiders p 6, 7, 24-25 Flying fish p 22

Two wings or four

Apart from birds, only two other groups of animals have developed wings for powered flight. They are the insects and the bats.

Insects have been flying around for a very long time. Fossils of giant dragonflies have been discovered which lived 300 million years ago.

▲ Insects like dragonflies use a pair of muscles to move each wing independently.

▼ As each wing flaps down, it pushes against the air. It is this pushing action that moves the insect upward and forward. Tiny rubber pads at the base of the wings make them bounce up or down after each beat.

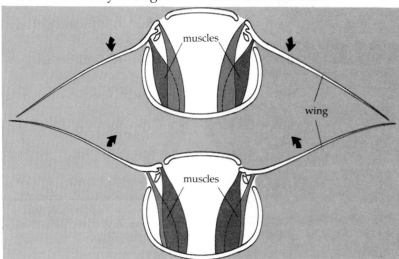

Dragonflies have two pairs of wings. They are very fast fliers, zooming over the surface of a pond at nearly 19 miles an hour (30 kilometers an hour).

Normally a dragonfly can control its four wings perfectly. But when it makes a sudden twist or turn the two pairs of wings sometimes rub together. On a still summer's day you can often hear the rustling noise this makes.

Up and down

Insects have powerful muscles which move their wings up and down. In insects like dragonflies, the muscles attach directly onto the base of each wing. But in bees and flies the flight muscles are arranged differently. In these insects, the middle part of the body, or thorax, is packed full of flight muscles. They make the thorax move in and out very quickly, rather like a tin lid being clicked in and out at high speed. In bees and flies it is these fast body movements which move the wings up and down and make them hum.

The wings of an insect are strengthened by small veins which support the thin membrane. When a butterfly hatches from a pupa, it pumps blood through these veins to make the wings unfold, ready for flight.

Two is best

Many insects find it easier to fly with two wings rather than four. And even if they have two pairs of wings, they often join front and back wings together to make just one flying surface.

Bees and wasps link front and back wings by rows of tiny hooks. Hawk moths are the "fighter planes" of the insect world. Their back wings are much smaller than the front pair and the two pairs are zipped up together by very small, bristly hairs.

Beetles do something very different. The front wings form a pair of horny shields which protect the "flying" wings underneath. This allows beetles to barge around in the undergrowth without damaging their flight wings.

▲ Butterflies and moths have four separate wings. The front pair overlap the ones behind. When a butterfly or moth flies, its four wings work together as one pair. Most butterflies and moths beat their wings about ten times a second. But they often rest by gliding between wing beats.

With knobs on!

Flies are probably the best fliers in the insect world. They have large front wings and small, knob-like back wings that look like tiny drumsticks. But the reduced back wings, called halteres, are important. They vibrate very quickly and work like miniature gyroscopes to stabilize the fly in the air. Some flies beat their wings as many as 1,000 times a second.

Warming up

The thorax of a bumble bee is packed full of flight muscles. But the muscles have to be warm to beat the wings properly. In cold weather, bees will shiver for several minutes before flying. It is their way of warming up their engines before take-off.

▲ A bumble bee has its own "fur coat" of tiny hairs which help keep it warm and ready for flight.

More about ⟫ Bird flight p 52-53 Insects p 24
Bats p 54-55 Dragonflies p 16 51

Bird flight

An ostrich weighs about 100 thousand times more than a hummingbird. Ostriches are not able to fly, but most other birds can.

Moving in air needs a special kind of body design. The body needs to be light in weight. It also needs wings to keep it airborne, and powerful muscles to move the wings. Flying birds have all these things. They are perfect flying machines.

A swift spends most of its life in the air. It can even sleep while on the wing. The swift's streamlined shape helps to make it one of the fastest animals. It often reaches speeds of 94 miles an hour (150 kilometers an hour).

Wing shape

A bird's wing is shaped like a shallow, upside-down saucer. It is curved on top and slightly hollowed out underneath. Because of its shape, the air pressure is always greater underneath the wing than on top. This gives the bird lift when it is flying and keeps it airborne.

▲ This gannet is landing feet first. Its large, webbed feet and fanned-out tail act as air brakes, helping it to slow down and land safely.

In a flap

Most birds fly by flapping their wings up and down. Large muscles provide the power. As the wings beat downwards, they push the air backwards. This makes the bird move forwards. When the wings are pulled up again, the feathers at the wing tips part to allow air to pass through. The wings are then ready to beat down again.

It is important to save energy when flying. Many small birds like starlings take a short rest between wing beats. They fold their wings against the body and move forwards like a ski jumper.

▼ Birds have large and very powerful chest muscles to beat their wings up and down.

swifts

Nature's helicopters

Hummingbirds fly like tiny helicopters. They can do almost anything in the air, including flying sideways, backwards and even upside down! Some hummingbirds beat their wings at more than 50 beats per second. This very fast movement allows them to hover in front of flowers while sucking the nectar from inside.

Gliding and soaring

Some birds hardly ever flap their wings. Instead they glide on outstretched wings for hours on end. Albatrosses have long, narrow wings for gliding on the strong winds which blow over the surface of the sea.

Other birds have become expert at soaring. Vultures and storks have wide wings which can make use of the invisible hot air currents rising from the land. Once a vulture has found one of these thermals it is carried up to a great height. If it slips out of a thermal, the vulture drifts downwards until it finds another.

▲ The wandering albatross has the biggest wings of any bird. It is an ocean glider and spends most of its life at sea on its 10-foot (three-meter) wide wings.

Flying hatchets

The hatchet fish is an unusual fish because it really can fly. It has a deep chest, just like a bird. This is where the powerful muscles are found which beat its wing-like fins up and down. It can fly for quite long distances just above the water.

The tinamou is probably the world's worst flyer! It cannot control its flying and often crashes soon after it takes off. It also gets tired and quickly becomes exhausted.

Tinamous have even been seen to dive suddenly into the water when trying to cross a river. However, they are good swimmers so they often reach the opposite bank.

More about ⟫ Ostrich p 38 Wings p 10-11, 50-51, 54-55
Gliding p 48-49 Flying fish p 22, 49

Furry fliers

There are nearly 1,000 different types of bats on the Earth's surface, and they form about one-quarter of all mammals. Apart from birds and insects, they are the only other main group of animals that can fly by flapping their wings up and down. They vary in size from the tiny kitti's hog-nosed bat (the world's smallest mammal) to the giant fruit-eating bats, which have a wing span of nearly 6.5 feet (two meters).

The wing is the thing

The wings of a bat are very different compared to those of a bird. The wing membrane is made of skin and muscle, and it is supported by arms, legs and extra long fingers. The membrane is stretched over the fingers like the skin over the ribs of an umbrella.

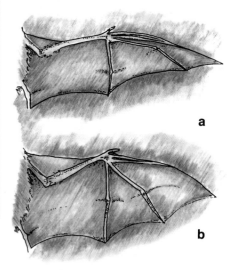

a

b

▲ Each wing of a bat contains the same bones as those in your own arm. But they are different in size and shape. The long, thin bones over which the wing is stretched are the bat's fingers. Can you pick out the bat's thumb?

▶ Bats with long, thin wings are fast fliers (a). They live in open country where there are few trees or other obstacles to get in their way. Wide wings (b) are better for slow flight. They allow a bat to maneuver more easily in thick vegetation.

Batting along

A bat uses its wings like a bird. The wings are moved up and down by large sets of muscles attached to its deep breastbone. During the downstroke, the wing surface pushes the air downwards and backwards. This keeps the bat airborne and moves it forward.

Bats have other adaptations to help them fly more efficiently. Many have short heads with a snub nose. This design makes the bat less heavy-headed and stops it from doing a nose dive once it is up in the air.

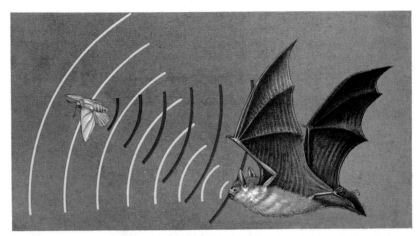

▲ Bats produce high-pitched sounds which bounce back off objects in their flight path. They listen to these echoes and use them to help find their way in the dark. This system is called echo-location and it works like a kind of radar. Bats also use their sonar system to catch insects.

▼ In most bats the wings are attached to the ankles. But in this bulldog bat from Mexico they join on to the knees. The bat is left with extra long legs armed with huge curved hooks for catching its favorite food, fish. It probably finds its prey by using its sonar system. But it needs remarkable flying skills to snatch a fish out of the water without crashing into the waves head first.

Hanging around the roost

Bats have two short back legs each ending in a foot with five clawed toes. The feet are used like a clothes hanger, allowing bats to hang upside down when resting. But some of the smaller bats also use their feet for moving around. They sometimes move on all fours, scurrying about on the ground on a mixture of feet and wings.

The common vampire bat has extra long legs and can even run and leap very quickly. When hanging upside down, bats even use their free thumbs to clamber about the roost.

More about ⟫ Bat wing p 11 Wings p 50-53
Sonar p 21

Animal olympics

50	60	100
80	90	100	160

The fastest and the slowest

It is not easy working out accurate speeds for some animals, especially the fastest moving ones. Some swifts can easily fly at 100 miles an hour (160 km an hour) in level flight but exact speeds are difficult to measure. The world record for any animal belongs to the peregrine falcon. In one experiment a falcon was timed at 217 miles an hour (350 km an hour) when making a very steep dive.

On land the cheetah is nature's sprint champion. It can run at 60 miles an hour (100 km an hour) and probably faster. But the cheetah lacks stamina and is easily beaten by some antelopes over longer distances. The fastest olympic sprinters fall well short of this. Their maximum speed is about 25 miles an hour (40 km an hour). Where would you fit the human on to this picture?

The streamlined sailfish wins in water. The specimen swimming at more than 60 miles an hour (100 km an hour) crashed straight through the side of a wooden boat!

But there are also nature's slow pokes. A sloth would not win a 100 yard (100 meter) race. Even when hurrying it would take nearly half an hour and a mole is even slower. Burrowing flat out it would need nearly 8 hours to cover the same distance!

Answers

Answers to questions in book

p 6: A hare.

p 6: Yes, your body shape does change when you walk or run. Your shadow shows this.

p 7: Yes, a snail does follow the two rules of movement. It pushes and it changes shape.

p 7: A spider pushes itself along on eight spokes, or legs.

p 8: The elephant's leg bones join on to the shoulder and hip bones.

p 9: When muscle A shortens the leg straightens. When muscle B shortens the leg bends.

p 9: He has to shorten or contract the opposite muscle, called the triceps.

p 11: They are the bat's finger bones.

p 22: The Bible describes Jesus as appearing to walk on the surface of the water.

p 30: An animal has to move or transfer its weight.

p 30: When standing, the animal's center of gravity falls in the area between its four legs.

p 34: A squirrel has a long body with short legs.

p 34: Yes, the model squirrel should be able to stand on a steeper slope than the horse.

p 35:
1. Both left feet are on the ground.
2. The right back foot is coming down towards the ground.
3. The left back foot is about to lift off the ground.
4. The body is supported by the left front and right back legs.
5. The right front foot is coming down and the left front foot is about to lift off.
6. The horse is now balanced on the two right legs.

p 35: When walking, a camel and a giraffe move both right legs and then both left legs alternately. This kind of movement is called pacing.

p 41: The snowshoe hare has large, flat feet like a pair of snowshoes to prevent it sinking in the snow.

p 54: The bat's thumb is the small pointed structure sticking out beyond the top of the wing membrane.

Glossary

air brakes: devices to help slow down an animal's speed. Birds use their tails and feet for braking when they land.

air resistance: something that prevents the smooth flow of air. A streamlined shape helps reduce air resistance.

aquatic: describes animals or plants which live in water.

ballooning: a technique used by spiders for floating in air. They use threads of silk like tiny parachutes.

bipedal walking: walking on two legs. Humans are bipedal and so are birds.

biceps: the large muscle in the front of the upper arm. It shortens to bend the arm at the elbow joint.

blubber: the fatty substance found under the skin of whales and seals. It is used for streamlining and for insulation.

brachiation: the technique used by apes, especially gibbons, for swinging through trees. The animal grasps branches with alternate hands as it moves.

buoyancy: the way some objects tend to float in water or air.

center of gravity: the point in a body around which its weight is evenly balanced or distributed.

chitin: a hard, horny, waterproof substance found in the outer covering or skeleton of insects and their relatives.

cilia: microscopic hairs on the surface of tiny animals living in water. They are used to paddle the animal along.

drag: the resistance an animal meets as it tries to move through air or water.

echolocation: a method of navigation using echoes. High-pitched sounds are sent out and these echo or bounce back off solid objects in their path.

flagellum: a long, tiny, whiplike structure found in some micro-organisms. It works like a propeller and drives the animal through the water.

fluke: the horizontal tail fin of a whale or dolphin.

friction: the force that slows down movement and produces heat. It is the drag effect when two surfaces are rubbed together.

gravity: the invisible force exerted by the Earth on all solids, liquids, and gases. The force of gravity on an animal's mass gives it weight.

halteres: the tiny "drumsticks" which are the modified back pair of wings of a fly. They work like miniature gyroscopes.

hydrofoils: a structure designed to give an animal lift in water. A shark uses its front fins like hydrofoils.

plankton: microscopic organisms that float near the surface of the sea and freshwater lakes.

prehensile tail: a tail used for gripping and holding on.

sonar: an echo-sounding system used by dolphins and whales for finding food and for navigation.

surface tension: a property of liquids that makes the surface appear to be covered by a thin, elastic film.

thermal: a column of warm air rising up from the Earth's surface.

Index

Acknowledgments

ARTISTS:

Dawn Brend; Adam Hook/Linden Artists; Mick Loates/Linden Artists; Alan Male/Linden Artists; Maurice Pledger/Linden Artists; Sallie Alane Reason; Diane Stutchberry; Helen Townson; David Webb/Linden Artists

Reference for Walking wheels (p.7) taken from *How Animals Move*, JAMES GRAY, Penguin Books

PHOTOGRAPHIC CREDITS:

t = top; b = bottom; c = centre; l = left; r = right.

COVER: Johnathon Scott/Seaphot. 6l Sue Earle/Seaphot. 6c A. Kerstitch/Seaphot. 6r Gordon Langsbury/Bruce Coleman Ltd. 7 Trevor Hill. 9 Sporting Pictures. 10tl Ken Lucas/Seaphot. 10tr Trevor Hill. 10bl Johnathon Scott/Seaphot. 10br John Shaw/NHPA. 11 Silvestris Gmbh/Frank Lane Picture Agency. 12 Peter Parks/Oxford Scientific Films. 13 D.J. Patterson/Seaphot. 14 Trevor Hill. 15 Jane Burton/Bruce Coleman Ltd. 16t Kenneth Lucas/Seaphot. 16b Rod Salm/Seaphot. 17 Survival Anglia. 19t, 19b Herwarth Voigtmann/Seaphot. 20 Menuhin/Seaphot. 21 David Rootes/Seaphot. 22 Nigel Dennis/NHPA. 23 H. Eisenbeis/Frank Lane Picture Agency. 24t Anthony Bannister/NHPA. 24b Rod Williams/Bruce Coleman Ltd. 25 Stephen Dalton/NHPA. 26 J. Cancalosi/ Bruce Coleman Ltd. 27 Jen and Des Bartlett/Survival Anglia. 28 Anthony Bannister/NHPA. 29 Carol Hughes/Bruce Coleman Ltd. 31 J.P. Scott/Seaphot. 32t Keith Scholty/Seaphot. 32b J. Robinson/NHPA. 33 Dani/I. Jeske/NHPA. 34 Gunter Ziesler/Bruce Coleman Ltd. 35 Harmut Jungius/Bruce Coleman Ltd. 36 Gunter Ziesler/Bruce Coleman Ltd. 38 Jen and Des Bartlett/Bruce Coleman Ltd. 40t S.C. Bisserôt/Nature Photographers Ltd. 40b Ken Lucas/Seaphot. 41l John Shaw/NHPA. 41r Anthony Bannister/NHPA. 43 Wolfgang Bayer/Bruce Coleman Ltd. 44 Kim Taylor/Bruce Coleman Ltd. 45t R.C. Hermes/Frank Lane Picture Agency. 45b Des Bartlett/Bruce Coleman Ltd. 46t John Visser/ Bruce Coleman Ltd. 46b A.N.T./NHPA. 47t Jany Sauvranet/NHPA. 47b G. Downey/Bruce Coleman Ltd. 48 Ivan Polunin/NHPA. 49 Jane Burton/ Bruce Coleman Ltd. 50 K. Wothe/Bruce Coleman Ltd. 51 Mik Dakin/Bruce Coleman Ltd. 52 Richard Matthews/Seaphot. 54 Stephen Dalton/NHPA.